# Projects: Making Hands-On Science Easy

*by*

Carl Tant

Biotech Publishing
A Division of
Plant Something Different, Inc.
Angleton, TX  USA

# Projects: Making Hands-On Science Easy

By Carl Tant

Published by Biotech Publishing
P.O. Box 1032
Angleton, TX 77516-1032

All rights reserved. No part of this book may be reproduced or transmitted in any form or by any means, electronic or mechanical, including photocopying, recording, or by any information storage and retrieval system without written permission from the author, except for the inclusion of brief quotations in a review.

Copyright © 1992 by Carl O. Tant

Printing 10 9 8 7 6 5 4 3 2 1

Library of Congress Catalog Number: 92-70485
ISBN 1-880319-06-3  $12.95 Softcover

**Publisher's Cataloging in Publication Data**

Tant, Carl
Projects: Making Hands On Science Easy
Bibliography, Index, Tables, Charts
1. Science projects
2. Science Education
3. Individualized Science

**Cover by Tammy K. Crask**

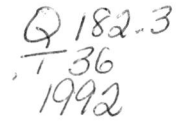

# Table Of Contents

**Chapter 1**  *Dream, Nightmare, Or Reality* ..... *1*

**Chapter 2**  *Science Project Objectives* ....... *5*

**Chapter 3**  *How Much Help?* ............. *9*

**Chapter 4**  *Attributes Of Science Projects* .... *14*

**Chapter 5**  *Project Evaluation* ............. *18*

**Chapter 6**  *"What Can I Do?"* ............ *24*

**Chapter 7**  *The School Or Classroom Project* . *33*

**Chapter 8**  *The Science Fair Project* ........ *37*

**Chapter 9**  *Let's Get Started* ............. *42*

**Chapter 10**  *Time and Money* ............. *47*

**Chapter 11**  *Let's Do A Project* ............ *56*

**Appendix A**  *Project Books* ............... *67*

          **B**  *Science Supply Sources* ....... *68*

## Dedication

*Following graduation, we look back upon our years as students, and, with some regret, realize how few really great teachers we had.*

*Edward Owen Bennett, Professor of Biology at the University of Houston, my graduate advisor, was one of that rare few. Those of us fortunate enough to come under his tutelage, tyranny--and care--have, I hope, multiplied his influence.*

*He could maintain rapt attention of a class through a three hour lecture, but more importantly, he could push, prod, and question with his students secure in the knowledge that his steady hand was there to rescue if needed. He taught us bacteriology, but, also, by example, he taught us how to teach.*

*To Ed Bennett, teacher and friend, this book is dedicated with a lifetime of appreciation.*

# Acknowledgement

A few of the many who contributed to this book with their ideas and suggestions deserve some special notice.

Suzanne Muecke, English Chair at Angleton Senior High School, used more of her endless red ink supply in proofreading.

George Seidel, Jr. of Seidel and Associates, initially brought the need for the book to my attention. Actually starting to write it was inspired in part by publishing consultant Joe Vitale's book, <u>Turbocharge Your Writing</u>!

Long overdue for thanks is Joyce Hawk, Account Executive with our printer, BookMasters, Inc. Her hand-holding, advice, and patience have gone far above and beyond the call. Undoubtedly, her accolade here deserves to be shared with Peggy Murray, Diana Fitzgerald, and many others in that organization.

## Warning--Disclaimer

This book is designed to provide information in regard to its subject. It is sold for the use of teachers and parents solely to serve as a guide and promote confidence that hands-on science projects do not have to be difficult. This background information is not by any means all that is known about the various subjects. Much of it is based on the author's personal experiences. Some might find this useful and directly applicable; others may not.

Because of the variables inherent in different environments and conditions for conducting experiments, their success can in no way be guaranteed. Any experiments are suggested here as guides only and should be adapted to fit individual situations. In every case, safety considerations should be emphasized.

As is common with any book, there may be mistakes both typographical and with reference to content.

The author and Biotech Publishing shall have neither liability nor responsibility to any person or entity for any loss or damage caused or alleged to be caused directly or indirectly by following the suggestions in this book.

**This book is sold with the understanding that the buyer will be bound by the above. If you cannot accept these terms of sale, the book may be returned to the publisher in saleable condition for a full refund.**

# Preface

Education fashions come and go. What is hot today suddenly experiences a chill tomorrow. Older procedures, thought safely laid to rest, suddenly become resurrected.

Like it or not, agree or disagree, we are seeing a resurgence of emphasis on hands-on education in science. It is catching many teachers unaware and ill prepared. The new requirements are being rapidly applied at all levels from elementary through high school. In some cases, as much as fifty per cent of the course time is required to be experimental investigative work. In the strict sense of "hands-on," computer simulations are not part of the picture. Many teachers face several dilemmas: How do you do it? How do you find time to do it? What do you do it with? Unfortunately, in many schools the increased laboratory requirement has been accompanied by reduced budgets. This is a stupid and seemingly unworkable mix, but the teacher probably was not consulted.

Despite such a state of affairs, the teacher is the one who will be held responsible and accountable. What is the way out?

The question brings us to the purpose of this book:

It is to try to take some of the stress out of requirements for individualized hands-on project work. The stress is not limited to the teacher or student. The assignment goes home, and parents become involved, particularly at the lower grades. Not knowing what is expected or what to do, the parent rightfully becomes stressed, also.

Teachers have enough trouble trying to keep up with the rapid advances in many scientific fields. This difficulty is particularly a problem in a field such as biology which has experienced unbelievable expansion in the last decade. If the teacher has been out of school more than about five years, much of the earlier knowledge is obsolete or only the basis for new technology. Heaven help the parents who may not have been back to school in fifteen years and have had no reason to keep up with the rapid changes. Let's face it. The science that students are studying today is not the same science that their parents learned. So how can the parents cope? How can the teacher cope?

This book is an effort to provide some answers--or at least an effort to provide a starting place for working out some answers.

The author enjoyed individualized project work during his years of public school teaching. As time went on, the problems surrounding such approaches became intensified with larger classes and budgets which did not keep up with cost increases. It is hoped that some of the experiences gained the hard way might be of benefit to those who carry on today. Much of what is suggested here is based on personal experience.

And that brings us to the next subject. The author and publisher decided at the outset that this book would be most useful if it were informal and user-friendly. Consequently, we have occasionally taken some liberties with the strictest interpretations of form. Even our hard-line grammarian agreed that in this case the use of the first person "I" or "We" would avoid awkward-to-read formal structure referring to personal experiences in the third person. If you do not agree with this, "Sorry, but that's the way it is." That is what both the publisher's grammarian and my secretary say when I try to sometimes split an infinitive.

The ideas suggested in this book worked for me, both as teacher and parent. They won't work for everybody. One of the great things about teaching and parenting is that different people accomplish the same objectives in their own different ways. Some of the procedures will be fine as written for many readers. For other readers, they will serve merely as starting points to stimulate thinking and modification for their own techniques and personalities.

Let's get this started with a freebie. At the risk of causing massive coronaries for the corporate attorneys, the author and publisher hereby grant permission for the users of this book to make a reasonable number of copies of the charts which they might want to use for their own students. This does not include permission to reproduce them in any other publication. It is restricted specifically to the charts on pages 23, 53, 54, and 55. We do this because realistically we know that some of you are going to copy it anyway, while others will request permission. This is an example of stress prevention. For those who would make illegal copies, it prevents worries of violating copyright laws. For those who would request permission, it saves us all a lot of paperwork because we would probably grant the permission requested. Preventing stress usually makes a lot more sense than having to deal with it when we are all up tight after it occurs. Hopefully, the ideas in this book will likewise accomplish that objective.

# Chapter 1

## Dream, Nightmare, Or Reality

*The melting remains of a late snow glistened in the sunlight striking the hillsides surrounding a group of neat brick buildings in the valley below. The first greening of spring was beginning to be evident in their well kept grounds.*

*Several strands of heavy barbed wire along its top belied the peaceful appearance of an ivy covered tall brick wall surrounding the buildings. A heavy wrought iron gate at the entrance was located under the sign:*

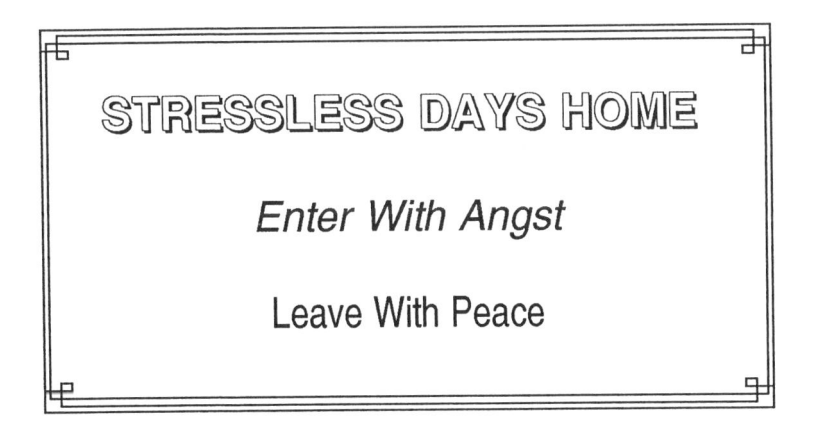

---

## STRESSLESS DAYS HOME

*Enter With Angst*

Leave With Peace

---

*In the entrance foyer of one of the larger buildings a group of young interns surrounded and listened to the words of the white haired chief of psychiatric services.*

*"At this time of the year we dedicate this building totally*

to the treatment of those suffering from a recently described condition--SPS. The initials stand for science project syndrome. The incidence of this condition is seasonally influenced and seems to be related to the requirements of many schools for students doing independent science projects. These occur most often in the late winter and early spring.

"Behind the door to my left is the section housing parents who have demonstrated tendencies of violence and threatened bodily harm to their children and teachers. To my right is the section housing teachers who have demonstrated similar tendencies of violence directed at students, parents, administrators, and even other teachers. The largest section located behind me houses both parents and teachers who are in various non-violent states. These range from simply sitting in the corner or staring at a wall mumbling meaningless gibberish, to some who go into a completely catatonic state."

A question was forming on the lips of one intern as he raised his hand for recognition, "What treatment can we offer for SPS?"

The chief shook his head slightly, "Unfortunately, we have no specific therapeutic agent. Fortunately, in most cases though, time and separation from the previous environment seem to be the great healers."

Another intern sought the senior physician's attention. "Do you mean there is nothing we can do to help these poor people?" she asked.

*The chief replied, "Oh, we can do the usual sedation and restraint when necessary to prevent the violent ones from harming themselves or others. Otherwise, a few months of rest and relaxation seem to be the best treatment. We are considering a new program of counseling this year. It will be directed toward enabling teachers to present the science project requirement without stress. Our past experience has taught that in many cases the stress results, at least partially, from the fact that teachers have had no training or experience in this field. We think that knowing more about it may help. We are looking forward to this program preventing relapses and return of the patients in future years."*

*"Is there any special way we should approach these patients?" asked another intern.*

*"No," replied the chief. "Just observe the standard procedures you were taught in medical school in dealing with any patient suffering from severe emotional distress and mental breakdown. We have found one thing which I do want to caution you about, though. That is to be very careful in talking with them to avoid using the words 'science' and 'project.' These two terms sometimes trigger intensification of their anxiety reactions."*

*The doctor signalled a guard to unlock the door behind him and led his charges in for their first contact with SPS patients.*

◆◆◆◆◆◆◆◆◆◆◆◆◆◆◆◆◆◆◆◆◆◆◆◆◆◆◆◆◆◆◆◆

Was that just one of those ridiculous bad dreams, or did it become nightmarish by approaching reality too closely? Have you ever felt the need for admission to a facility like the one described? If so, you have much company. Probably most teachers and parents have at least once felt that way when confronted with science projects.

Hopefully, even as more independent investigative work is required at all levels by new science curricula, this book will help alleviate the problems before they begin. Most of it is based on the author's almost quarter century involvement in science projects as teacher, parent, and science fair judge. After a few false starts, he figured out a way to handle it without the stress. In fact, believe it or not, science project work became one of his most delightful experiences both as a teacher and parent. Maybe some of these ideas will work for you.

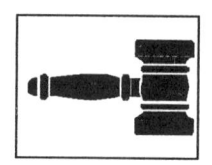

*This symbol is used throughout the rest of the book to emphasize especially important points.*

# Chapter 2

## Science Project Objectives

The primary purpose of science projects is not to teach science.

# WHAAAAT?!!

No, that first statement was not a big ugly hairy slimy typo. Nor has the author taken leave of his senses at this early point in the book.

Learning some scientific facts or even principles is a very valuable fringe benefit to students doing projects. The primary objective in properly conducted science project work is to teach them to **think**.

That huge little five letter word encompasses most of the major purposes of education:

>            **learning**
>            **planning**
>           **organization**
>          **interpretation**
>           **measurement**
>        **validation of information**
>  **relationships between different areas**
>       **communication of knowledge**

Adapted to any grade level, all of those goals can be

accomplished by strict adherence to what has become known as scientific methods of research. You have all probably heard them more than once, but they are vitally important. Failure to follow them in past centuries resulted in such silliness as the idea that if a pregnant woman slept on her right side, the baby would be a boy; if she slept on her left side, it would be a girl. These principles are important enough to state and restate, iterate and reiterate, so here goes:

> 1. Decide upon an hypothesis as a possible answer to the question of the subject.
>
> 2. Collect data. This means review the literature and find out what is already known about the subject.
>
> 3. Design an experiment to prove or disprove the hypothesis.
>
> 4. Observe results.
>
> 5. Repeat the experiment.
>
> 6. If further repetition provides similar results, the hypothesis can tentatively be interpreted as valid. If the experimental data does not support the hypothesis, a new hypothesis should be determined.

Adapt the goals and objectives to the age and ability level of your students. This can be highly variable. You are the expert in making that decision, if for no other reason, simply by virtue of the fact that you are there with the student.

## Divide to Multiply Success

Students who are not incorrigible problems in every respect will respond favorably if they meet success along the way to completion of their project. The final goal and hoped-for result should be kept in mind, but there are many intermediate steps to get the satisfaction of accomplishment along the way. Again, this will vary with the student's situation. For an elementary student, it might involve learning to use a meter stick; for an advanced high school senior, it might involve learning to interpret the print-out of a gas chromatograph.

## Science Is Not Isolated

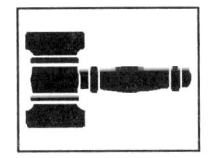

Doing a project is a fantastic way for students to learn that *science does not exist as an isolated island alone.* They should be encouraged to relate their subject to its history and social implications. Applications of mathematics can find use in most projects. Certainly of great importance is development of the skills necessary to effectively communicate the research results to others. This area needs emphasis whether your student has high scientific aptitude or not. A quick survey of current scientific journals will reveal that, unfortunately, many of todays leading scientists are not adept at communication.

As important and desirable as these goals are, they must remain subsidiary to the primary objective in doing science projects.

## *What You Really Teach*

Permit me to relate a story. It is a true story of a boy and girl who were outstanding students during the midyears of my teaching career. Several years after high school graduation this girl returned for a visit. In the course of our conversation, she remarked that she and her classmate had finally figured out what I had taught them. With some trepidation, I inquired what that was.

She replied, "You taught us a lot of biology and chemistry. You taught us a lot of techniques. But those were not important. Oh, all the facts and so forth that we learned helped a lot in college, but what you really taught us is how to think."

Could any teacher ever hope to receive a higher compliment?

# Chapter 3

## How Much Help

This is probably as good a place as any to lay to rest the perennial nagging question of many parents and teachers.

*"How much help can I give my student with this science project?"*

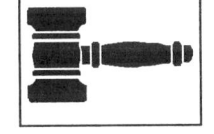

You may teach methods.

You may suggest how to go about finding information.

You may demonstrate techniques where manipulations are involved.

You may give physical assistance where appropriate as in handling a heavy or bulky object.

You **must** point out safety considerations.

You may **not** do the work.

### *Yeses and No-No's*

For example, a young student needs to make a poster to display the project results. It would be legitimate for the parent or teacher to show the student how to use a straight-edge and make a lightly penciled line to use in aligning lettering. It would not be legitimate for the parent or teacher to actually put the letters in place.

Another example: the student needs to use some unfamiliar literature to gather data. It would be proper to show the student how to use that publication's indexing system. You will be going too far if you actually look up the subject topics for the student.

From the middle elementary years on, many students who take pride in their work would like to present neatly typed reports and data. Certainly, such an attitude should be encouraged. However, many of the younger students have not had the opportunity to study typing or even learn proper keyboarding. In such cases, it is certainly a proper function for parents or friends to type the material. This provides an opportunity for the student to learn the appropriate form of expressing appreciation in an acknowledgement appended to the presentation.

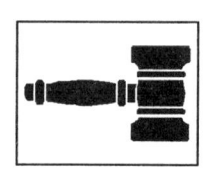

The most valuable help that teachers and parents can give is often indirect and almost intangible. It involves leading the student to the proper road. *Sometimes the most difficult part of the leading process is to stand back, be quiet, and let the student do the work.* In some cases, this involves making mistakes. Your function as the adult supervisor is to help prevent the mistakes from being too serious and discouraging. Help the students develop the confidence to know that although they are led to the edge of the cliff, you are there to stretch out a rescuing hand if necessary to prevent them from falling off the edge. If this confidence and attitude exists, the student will be more able to explore and accomplish.

## Let's Get Acquainted

One major benefit of science project assignments is that they provide teachers and parents an opportunity to really become acquainted or extend acquaintance with their students. The project provides a focal point for talking with the student. Note the phrase, "talking with." That means conversation. It means two-way communications. It does not mean issuing orders and instructions. Conversation implies listening.

Irrespective of their age, background, or interests most students feel at least some degree of pressure when they receive a major assignment such as doing a science project. They have numerous concerns. Some will express these; others will not. Conversation with the student provides the means to bring their worries to the surface so that the adult can deal with them.

## Project Selection

A major hangup for most students is in the initial selection of a subject for the project. This is where you can provide much needed help. Talk with the students to find out what their major interests are if you do not already know. You can then point them toward different aspects of those interests which could be utilized in project work. One good approach is to suggest several subjects as discussed in detail in Chapter 5. *Above all, except as a very last resort, avoid assigning a specific topic.* There is nothing worse than having to spend much time and effort doing something which is not interesting. That approach will not lead to a good project.

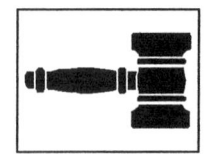

## Plan and Organize

Another extremely important need for help lies in the area of planning and organization of the work to be done. Unless the student has had several years of experience with project work, he or she will probably be grateful for help in outlining and organizing the approach to be used. Actually, this involves what scientists call a protocol, which is simply a statement of the procedures to be used. Protocols should not be intimidating. They are no more than a type of listing of the steps to be followed. They are not cast in stone--they may be changed as required by progression of the work. You will find more details about this in the next chapter. Two forms of protocol are given as part of the experiment in Chapter 11.

## Keep It On Track

A final note of forewarning is in order at this point. The younger and less experienced the students are, the more tendency they will have to try to go in too many directions at one time, taking off on tangents to the primary subject procedures. Your guidance here will be invaluable. This is a place to exercise caution not to discourage further interest. One good approach when the student seems to be on the verge of getting sidetracked is to observe that the student's idea is great--in fact, so great it might be a subject for another project or an expansion of the present one next year. Most students will react positively to these suggestions.

# Chapter 4

## Attributes Of Science Projects

All well-done science projects, whether conducted by an average third grader or a scientifically talented high school senior, should have certain basic characteristics in common. These include:

1. **Statement of purpose**. The statement of purpose usually includes the question and hypothesis which is the basis of the research.

2. **Background information**. This is a brief justification for starting at the beginning point of a specific project. Sometimes it can be incorporated into the statement of purpose or discussion sections. In such cases it is not necessary to list it as a separate item.

3. **Procedure**. Initially this will be the protocol. Modifications made to it as the work progresses should be included.

4. **Results**. This is simply the data obtained in doing the research work.

5. **Discussion of results**. Here is the place to interpret the data obtained. Did it answer the initial question "yes" or "no?" If an answer was not obtained, the discussion is the proper place to explain and give possible reasons why. A "negative" result should be

interpreted in a positive manner. Knowing what does not work is often a necessary prelude to finding what does. Be prepared - selling this idea to highly motivated students can be a difficult task.

6. **Future work.** The student should briefly describe the next or related logical procedures which might follow the present project.

7. **References.** Basically this is a bibliography of materials studied by the students in preparation for the project.

8. **Acknowledgement of assistance.** Very little research, even at the professional level, is done by a scientist working alone. Usually, there is an input of suggestions or criticism from others if more tangible specific help is not provided. The sources of all help should be listed and identified with the type of assistance given.

## *How Much Is Enough?*

Students will be concerned about how extensive each of the items above should be as they are assembled into the project report. There is no single simple answer to that question. It will vary depending upon the nature of the subject and research done. Consequently, in most cases it would be a serious mistake on the part of the supervising adult to set minimum lengths on each section. Some subjects will require more in particular areas than others. Each topic will have to be judged on background information available and the specific procedures used.

Again, keep in mind where you are. A sophisticated three or four year continuing project of a high school student might have two or three hundred references accompanying an extensive literature review. Obviously, this would not be appropriate nor expected for an elementary student.

The other topic of greatest concern to many students is how detailed the procedure should be. That question can often best be answered by suggesting to the student that the steps described should be given in such a way that someone who knows nothing about them can follow the procedure and do the exact same thing that was done in the research.

This is a golden opportunity to introduce the student to a harsh lesson most adults have experienced at least once:

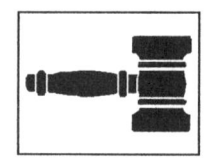

*Don't take for granted that somebody else knows what you are talking about.*

## What Happened to a Simple Instruction

The step in the procedure states "clean the test tube." The experienced student knows what to do and proceeds. The beginner may ask:

"How?"

"Do you wipe it out with a dry cloth?

## Attributes Of Science Projects

"Do you simply rinse it out with tap water?

"Do you use a test tube brush with a detergent solution?

"Does the tube need to soak over night in a cleaning solution to remove gummy material?

"Should it have a final rinse in distilled water?

"Should it be dried by wiping or simply by being inverted to drain?"

It is obvious that the factors underlying the question addressed here are not limited to science projects alone; they apply to many aspects of life. That, after all, is one of things requiring science projects is all about.

# Chapter 5

## Project Evaluation

Teachers as well as students and their parents frequently agonize over how a project will be graded. The grade stress can be greatly reduced if students are made aware at the outset how their project will be evaluated. The following are major factors to be considered in reaching a final grade. The weight assigned and extent of each may have to be adjusted for different age and background levels of the students.

1. **Scientific thought.** Does the student evidence logical thinking and procedures in various aspects of the project?

2. **Thoroughness.** Did the student's work cover all major areas of concern? Was testing carried out to a reasonable degree of completion? Did the student redo or repeat procedures which experienced problems? Did the student adequately cover the pertinent literature references available?

3. **Appropriate conclusions.** Are the conclusions drawn supported by the experimental results?

4. **Presentation of the project.** Included here would be appropriate and attractive display, clarity in communicating to others, and completeness. Depending upon the type of presentation required, this area could be subdivided to give separate grades

for oral and visual presentation.

5. **Goal achievement**. Did the student either succeed in answering the initial question or explain why the results did not answer it? The latter should have value equivalent to that of a positive "yes" answer.

6. **Interim reports**. If you require progress reports during the course of the project, these should be given some value in the final overall grade.

The last item above, interim reports, can provide much valuable information for the teacher and help the student keep the project on schedule. Many parents like to have a list of the reports that will be required and dates when they become due.

## *Reports & Deadlines*

1. **Initial report.** This should be no more than the name of the subject to be investigated. The title does not have to be in final form. The teacher might require the student to write a sentence or two briefly stating why that topic was selected.

2. **Reference report.** This would become due a reasonable time from the date for the topic selection. It does not need to be elaborate. Simply require a listing of a few references read.

3. **Research progress.** At some midpoint between the initial reports and the final completion date, students

should be required to submit a written summary of what has been done up to that time. Depending upon grade level and other local factors, this type of report could be required more than once. Facing a deadline to have something done will do much to head off an attempt to do everything at the last minute.

4. **Rough draft** of the final report. It might be a good idea to simply give a check grade for having this in at the proper time. A letter or numeric grade on the initial rough draft is of questionable value. After all, we are talking about a rough draft. This is simply a means of helping the student organize the results. It gives the teacher and, in many cases, parents a basis for guiding the student into the selection of a desirable final format.

To obtain the greatest value from this, the student must be aware of dates which will be deadlines for each phase of the work. Many parents will appreciate having a copy of the schedule. Table 5-1 is an example.

## Grades and Records

Keeping records is a matter of teacher preference and individual school procedure. The interim grades could simply be included in the overall grade book. Many teachers find it worthwhile to have individual check lists. Figure 5.2 shows an example of such a list.

The weight assigned the project for a particular grading period should be related to the amount of time and work expended. It will not necessarily be the same in all schools, nor will it be the same at different grade levels. It should--in fact, must-- however, be the same as used by each teacher of the same grade level in a particular school. This means that the teachers involved need to get together and assign specific values to the different phases of the work.

## Table 5-1

## Science Project Schedule Example

| | |
|---|---|
| January 5 | Announce project to class. Send brief explanation and time schedule to parents. Make general explanation of what is expected to students. |
| January 7 | Discuss project subject selection with students. |
| January 8 | Provide materials to help students select subject. |
| January 9-15 | Provide time for further work on subject selection. |
| January 17 | First report due -- Title of project. |
| January 19-24 | Provide time for reading about subject. |
| January 27 | Second report due -- List references. |
| January 28-February 10 | Working time. |
| February 11 | Third report due -- Research progress. |
| March 1 | Research completion. |
| March 5 | Fourth report due -- Rough draft of final report. |
| March 6-13 | Correct and discuss rough drafts. |
| March 20 | Projects presentation. |

Project Evaluation

## Figure 5.2

## Interim Science Project Grade

Student _____

Report No. _____  Date _____

Project Subject

_____
_____
_____

|  | Excellent | Good | Average | Poor | Unaccept-able |
|---|---|---|---|---|---|
| Procedure thought |  |  |  |  |  |
| Organization |  |  |  |  |  |
| Thoroughness |  |  |  |  |  |
| Research Progress |  |  |  |  |  |
| Records |  |  |  |  |  |
| Appropriate Conclusions |  |  |  |  |  |
| Presentation Progress |  |  |  |  |  |

# Chapter 6

## "What Can I Do"

Many students are very sincere when they say, "I'd like to do a project, but I don't know what."

This is a major stumbling block for students at all levels. Relatively few, at least initially, have a clear idea of some subject they would like to investigate. The problem is inversely proportional to the student's age and experience. Inherent in the question is an underlying fear that the project subject must be very involved, difficult, and sophisticated. *Students need to be reassured that such is not the case.*

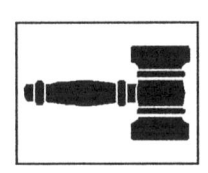

Many middle and high school students, especially advanced ones, add the further complication of feeling that a project must be original research. In other words, they feel that for it to have any value it must be something that has never been done before. This is certainly a commendable attitude, but not a very practical one. It would apply only to a very few students who are looking for an extended continuing three or four year project which they intend to exhibit at an international science fair by their senior year.

Doing something different is praiseworthy, but being different does not imply difficult. Even cookbook projects, such as many lab manual experiments are, can be used as a starting point for further investigation. At the lower grade levels it is certainly appropriate for the student to learn to follow specific steps. This is a worthwhile goal in itself at that age. Even here, however, the student could be encouraged to extend at least one phase of the project. A very successful project could be no more than simply carrying out an experiment precisely as outlined and comparing the results obtained with those which occur from utilizing the same basic process, but with a few variations. The variations can be quite simple, for example, a different type of container, different light, different temperature, etc.

## Sources of Ideas

The rest of this chapter is devoted to sources and examples of ideas that might be presented to students. *There are few instances, if any, in which a specific topic should be assigned to individual students.* There is nothing worse than spending time and effort working on something in which the student really has no interest, or perhaps even dislikes. That will neither achieve a good result nor will it accomplish any educational objectives. Give them a choice.

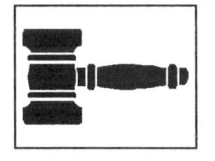

A good source of ideas is laboratory manuals. Many of the experiments can be used essentially as described or can be easily expanded. Most lab manuals are written for a particular grade level. For students of average or above ability, it might be worthwhile to let them look through the experiments in manuals a grade or two above their own level. This will give the challenge that many are seeking. Some students will want to make up their own variations of the procedures or goals described.

Go to the library. Most libraries will have at least a few books on science projects in different subject areas. A note of caution is in order here: unfortunately many of the books may be several years old and will describe experiments which should not be done because of modern safety requirements. The appendix lists some recent books.

Check the reference section. Many encyclopedias, especially those for children, such as **Child Craft**, have excellent ideas for project work or investigations which can be adapted. Other good sources would include books on many different subjects for children or young adults. Frequently, the discussions of scientific subjects will bring out areas which might be suitable for investigation.

Current science periodicals or general periodicals with science sections written for different grade levels are excellent sources of good ideas.

Watch TV. Yes, you read that correctly. There are a few good programs. Some, like Nova, *can be a bountiful source of ideas.* Even some of the older ones, such as reruns of Mr. Wizard, also present ideas which can be used or expanded upon.

Collections can be valid projects, especially for younger students. The collection, however, must be more than simply a box of rocks, insects, or leaves. Part of the research involved with a collection would be proper identification and classification of its components.

## Science From Non-Science

Look around and begin to develop an awareness of common things which could be turned into very interesting project work. For example, a trip to the grocery store could easily turn up a number of projects. One might involve a comparison of nutritional content of competing brands of the same food. Another subject which has often aroused the author's curiosity is how realistic are the serving sizes listed on the labels. This might even be extended into a comparison of realism with respect to foods which are basically "good for you" with those which contain high quantities of fat, cholesterol, or sugars. Another one might be a comparison of preservatives. This could go on and on.

A trip around the yard or to a park will yield many other ideas.

Some subjects which on the surface seem so simple that they are often overlooked can be the basis for excellent projects. For example, compare some different brands of canned or frozen food with respect to water content. Simply take the initial weight as listed and compare it with the weight left after all the water has been cooked away. This one might be a good eye opener for your pocketbook as well as a good science project!

## Notebooks are for Adults, Too

Develop the habit of making lists or jotting ideas down on note cards. *Do it as they occur to you--these things have a*

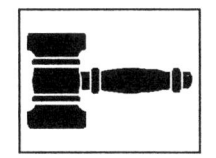

*tendency to escape the next day.* After a year or two of this, your collection of ideas will become so large that you will want to start thinking about putting it into some type of computer data base. At the middle and high school levels, another good source of ideas is a listing or abstracts of previous year's science fair projects. Such publications are obtainable from most regional science fairs affiliated with the International Science Fair. A similar source of ideas is the listing of winning projects in the Westinghouse Talent Search.

One of the best sources of ideas is the students themselves, who often are simply not aware that they have them. Lead them into thinking about something they would like to do by asking questions relative to their likes and interests. This could include such things as "What have you seen on TV lately that you liked or wondered about?"

## Get Something from your Taxes

Another good source of ideas will be found in numerous government publications. NASA publications and some put out by the Department of Agriculture are a particularly rich source of not only ideas, but also in some cases, specific detailed project plans. Publications such as these are obtainable at the government printing office outlets or are available in many libraries. Various pamphlets and booklets available from local county agricultural extension or home demonstration agents are valuable resources.

## Browse the Catalogs

Most teachers have general school supply and laboratory supply catalogs readily available. A quick look through the sections designated for your grade level will reveal numerous laboratory kits. While most of these are designed for classroom use, the subjects can provide good ideas. Students will enjoy looking through these to find something which interests them. If you do not need the classroom quantity, many of the ideas can be done on an individual basis. Some of the kits such as those put out by SYNTHEPHYTES™ have teacher manuals which include lists and suggestions for extension of the standard experiment into individualized project work. A fringe benefit of letting students look through these supply catalogs is that many will become aware for the first time of the costs involved in the educational supplies which they use.

Read the ordering instructions carefully. Some companies encourage individual orders while others are not set up to handle them.

## A Fun Example

Let's take a specific example of how a simple project can provide ideas for other experiments of varying complexity. One experiment for elementary students in the author's book, **Seeds, etc....**, is designed to show the effects of light on the production of the green pigment, chlorophyll, in most plants. The experiment utilizes a sweet potato vine but would work equally well with many other kinds of plants. Here are the instructions as given in Chapter 7, "Having a Vine Time."

"Cut out some figure, perhaps your initials, from black construction paper and securely tape it to the upper surface of the leaf. After about a week, remove the figure from the leaf and observe the effects of lack of light on leaf color. Be sure to select or arrange the plant so that the leaves will remain in a constant position with the upper surface clearly exposed to the brightest possible light source. The experimental set up for this is shown in Figure 7.3.

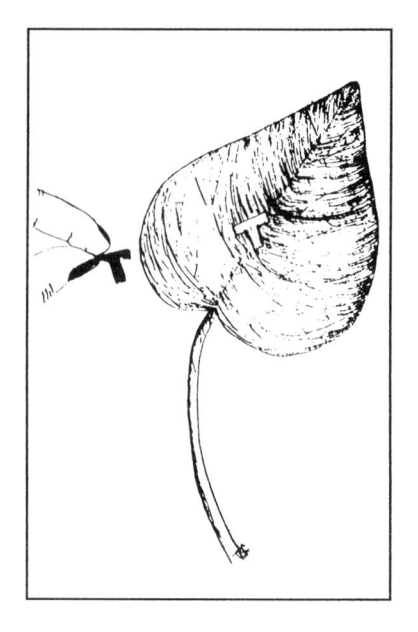

**Figure 7.3**

"Resist temptation to glue the letters or figure to the leaf. Most school glues are water soluble, and the material may fall off as a result of transpiration of water by the plant. Some glues contain substances which may be toxic to the plant tissue and result in a ruined experiment and injured plant."

At this point, I am going to revert to acting like a teacher. Please bear with me. Stop reading at the end of this paragraph, get a pencil and paper, and go back through the experiment described. Think about and list ideas which come to mind for extending it.

Now, see how your list compares with the one below. You probably thought of some things I didn't and *vice versa*.

1. What effect, if any, would be obtained if the letter were placed on the lower side instead of the top of the leaf?

2. Would the effect be identical the same time period if the test were conducted simultaneously on an old and young leaf on the same plant?

3. How would the effect compare when different types of plants are used? This in itself could be two different ideas. One might compare the effects of the test plant with other herbaceous plants. Another would be to compare the effects between herbaceous and woody plants. Still another might be to study the results and compare house plants with outside plants.

4. What percentage of the leaf surface could be covered before harmful effects, perhaps even death, are noted? You could further extend this idea by going back to number three and applying it under the conditions noted there!

5. Now, let's promote this basic third of fourth grade project to high school. If equipment is available (and it doesn't take much), the basic procedure could become a complex science fair project involving measurement of carbon dioxide uptake

and oxygen release as the partially covered leaf is compared with an uncovered one. The processes involved could be further extended with a variety of statistical analyses and studies of the effects of photosynthesis on other physiological processes, not only of the leaf, but also of the entire plant.

6. Is the process demonstrated limited to green plants or would similar results be obtained if the letter were applied to leaves of other colors such as might be found on a coleus or caladium? If a similar result is obtained on these plants with other photosynthetic pigments, would the time required be the same as with green plants?

This could go on forever, couldn't it? Since one of the underlying ideas of this book is that it should be short enough to be quickly and easily read by teachers and parents, it is time to stop with one final thought.

7. Developing ideas to expand a topic could be a good idea itself. Turn this approach into a fun brainstorming time for family or class. Pick a simple experiment like the one described, set a time limit, and see how many different ideas for expanding it the group can think of.

Could you justify number seven above as hands-on instead of a heads-on research if the students write their ideas down? Hmm.

# Chapter 7

## The School or Classroom Project

Science projects which are done simply as part of a course curriculum are the easy ones. They may be limited to a single subject of a single teacher, or they may result from a school requirement. In many cases, they can be thought of as little more than a regular lab procedure. The difference is that instead of all the students doing the same thing, each is working independently on a different subject. With this type of project, most choices and requirements are left up to the individual teacher. If the requirement is one for all science classes, it might be worthwhile for the science teachers to get together and establish some ground rules which could be followed by all. Doing so will avoid comparison problems by students who have different teachers. Who needs the headache of "Why do we have to do this? Mrs. So-and-so's class only has to do..." Some degree of standardization works wonders here, relieves student feelings, assures a sense of fairness, and makes life easier for all concerned.

One of the earlier decisions to be made will resolve the question of whether all work is to be done at school or parts of it at home. If school is the site of all the work, there will be little need for parent involvement. With such an approach, the role of the parent would be no more than with any other major assignment. If a large part of the work, particularly the experimental phases, is to be done at home, the parents should be made aware of all requirements, time schedules, etc.

Another decision relates to whether the project is individually done, done by pairs, or done by small groups of students working together. You can quickly think of many pros and cons for each approach. They are no different when applied to science projects than when applied to any other subject.

Students should be informed at the outset about the scope of project expected. They need to know not only the extent of work required, but also they should be made aware of how the project will be graded and what percentage of their grade it will count. For most classroom projects, the grade should not be so great during a single grading period that is could result in failure for that time. Whatever the decision is with regard to the weight assigned the grade, in all courses students should be made well aware of it in advance. Requirements for presentation constitute another area in which teachers should get their acts together if the science project is required in more than one class. For example, will presentations of the project be oral, or as poster displays, or as written reports? Some students will resent presenting their work orally to the class; others will resent it if they can't! If all students in the school have to abide by the same rules, a sense of fairness will exist.

One aspect of the project which is not a subject for variation is that of safety. All school safety regulations should be strictly enforced. The students should be aware of their importance whether they are doing the work at school or home. If the research is being done at home, the parents have an opportunity to reinforce the safety procedures taught at school.

Regardless of their personal feelings about animal experimentation, teachers would be well advised to limit experiments with vertebrate animals to those of an observational type related to pets or animals in their natural environment. This can avoid many problems.

Even though the classroom projects are not destined for an official science fair, you will find the following list of safety rules and suggestions a convenient one to use. Another thought for parents and teachers to keep in mind is that quite often a student will become so involved in a classroom project that he or she will want to extend the project into a full scale one for a science fair entry. This, of course, would apply only to students in eligible grades. However, many successful science fair projects in later years had their origin in the interests students developed in classroom projects of the elementary grades.

## *Safety First*

One of the early steps in any project assignment is a thorough review of safety rules. If project work is to be done at home, parents should receive a copy of the safety regulations.

The starting point is always the local school rules. Following are some items which need special emphasis.

1. No toxic or carcinogenic chemicals should be used. Your school should have a list, or you can obtain one from your state education agency.

2. Eye protection should be used with all chemical reactions, heating procedures, and mechanical processes which might produce flying particles.

3. Exercise special caution in projects involving insects.

4. Only commercially produced electrical equipment bearing the U.L. symbol should be used with all devices exceeding low voltage battery power.

5. Projects involving the growth of molds or bacteria should not be done except under direct professional supervision. It is too easy to accidentally culture an unrecognized pathogen.

6. Projects involving human body fluids should be prohibited. This certainly includes blood and saliva. Don't encourage the spread of AIDS.

7. Prohibit or discourage experiments with vertebrate animals. Many considerations other than safety are involved in animal experimentation. Strictly observational projects of animals in their natural habitat are usually all right, but observe safety precautions.

Now, let's move on and take a look at science projects for science fairs.

# Chapter 8

# The Science Fair Project

The term "science fair" brings to mind the large extensive fairs operating as part of the International Science Fair under the sponsorship of the National Science Foundation in the United States. Similar government sponsorship is found in other countries. These, however, are not the only science fairs. Many others are carried out on a local basis. Such fairs are more like group exhibitions of classroom or school projects. Students do not progress from this type of local fair to the International. They are, however, a good source of experience and may serve as the starting point to generate further interest.

If your student is within the eligible grade range for the International Science Fair or its affiliated regions, participation in a local non-affiliated fair is still possible. However, teachers and parents should make certain that the projects follow the international rules. Quite often the local rules do not encompass everything prescribed by the international. Consequently, if there is a rule conflict, the student may be ruled ineligible at the affiliated international fair.

It is logical to assume that parents who are reading this book have a high degree of interest in their children's welfare and progress. A word of caution is in order. Give careful attention to Chapter 3, "How Much Help?" Judging at the science fair involves questioning of the students. One of the judges' assigned tasks is to ascertain by questioning if the student received excessive adult help. If the decision is that the student did not do all the work, the result will be a poor score and perhaps embarrassment for all concerned. Be sure that all help given is properly documented in the "Acknowledgement of Assistance." There will rarely be any problem if the help is listed. It is the hidden help that judges delve for.

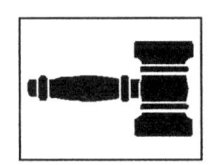

Science fairs have numerous forms to be completed at different dates and stages of the investigation. *One of the most important functions for parents and teachers is to be sure that these are completed properly by the specified deadline.* Failing to do so will result in the student's project being declared ineligible at the opening of the fair.

## Category Selection

Placement of the project in the proper competitive category in the science fair is another matter of great concern to teachers. Obviously, some projects could be entered in only a single place. A quick glance through the projects in this book will show that many others could legitimately be directed toward several areas. This subject should be considered months ahead of science fair, particularly if your school will have many entries and the number per category is limited. Emphasis on different areas will permit you to have more students. Do not stretch categories too far, however, as a project will not do well in judging if it is misplaced. Again, consult the student. Discuss the options available and the pros and cons of each. Then, insofar as possible, let the student make the final decision. A key point for consideration by both teacher and student is the question of area in which the student feels most confident for discussion of related subjects with the judges. If scoring is close, a student's knowledge of related areas may become a critical factor in final award placement. Table 8.1 shows the most common categories in large science fairs. The standard abbreviations will appear on many forms.

## Table 8.1

| | |
|---|---|
| BSS | Behavioral/ Social Science |
| BIO | Biochemistry/ Microbiology |
| BOT | Botany |
| CHE | Chemistry |
| ENG | Engineering |
| ENV | Environmental Science |
| ESS | Earth/ Space Science |
| MED | Medicine/ Health Science |
| MTH | Mathematics/ Computer Science |
| PHY | Physics |
| ZOO | Zoology |

You can avoid paper work problems by requiring that all permission forms, registration forms, protocols, etc. be in your hands at least two days before science fair day. That will give you time to review them for proper completion and signatures.

If space permits, set a deadline for completion of projects and bringing them to school a week ahead of the science fair date. During the month before this, repeat many times your usual speech about planning ahead. This will get about half of the projects there on the date specified. On that date, repeat your speech, emphasizing it with desk thumping. That will get most of the remaining projects there by three days before science fair. Then, be prepared to take whatever action you think appropriate to deal with the one or two projects which cannot possibly be completed ahead of time because of a last minute major discovery or disaster.

## Some Final Thoughts

A compelling reason for requiring early completion of projects is to make certain that related disciplines have received proper attention. It may be difficult to convince some students that all the scientific knowledge in the world is valueless unless it can be communicated. Insist that reports and displays be grammatically correct. Double check for spelling errors. Make certain that the conventions for italics and capitalization in scientific names of organisms have been properly followed.

Check any mathematical applications. Be sure that

statistical analyses have been appropriately selected for the type of data utilized. As preparation progresses, you may find it worthwhile to enlist the cooperation of English, math, and art teachers.

During the year as projects begin to develop, insist that proper records be kept, preferably in a bound notebook with each page properly dated and signed. Some students will want to recopy their laboratory notebook to make it neater. *Don't.* This is the original record of work that was done and should be presented as such. Certainly, the notes should be legible and accurate, but judges will be suspicious (and properly so) of what they suspect might be deliberate preparations that do not really have the appearance of having been done while the work was conducted.

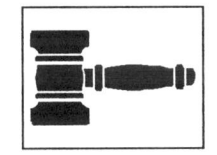

# Chapter 9

## Let's Get Started

This chapter is really for beginners at the science project game. Once you have done it a time or two, you will have learned, perhaps the hard way, how to get organized and get the projects underway.

For some, the first step is an attitude adjustment hour.

**Smile.**

Once you get into project work you will find that the event is much easier than the dreading and fear of it. Think positively about it. If necessary, make a list of the good things from humorous moments to valued learning experiences to career awareness that can result from students doing projects. If you compile a similar list of problems associated with projects, you will find that it will be much shorter. Many, if not most, of those problems can be eliminated before they start by simply planning ahead and knowing where the rocky places are in the road.

The next step is to develop a positive attitude in the students. For many students, this will be no problem because they will look forward to project activity with eager anticipation. Others will respond to the attitude of teachers and parents. If you think students cannot detect an adult's negative attitude, you have no business being either a teacher or parent.

## Let's Get Started 43

Much of the hassle, frustration, and stress associated with science projects results quite simply from waiting too long to carry out the various aspects of the work. This can be eliminated to begin with by proper planning and establishing a schedule. Dates should be set for at least the following stages and, as discussed previously, interim grades should be given.

1. Subject selection.

2. Review of pertinent literature.

3. Plan for conducting research.

4. A report on the initial stages of the research after it is under way.

5. Completion of research.

6. Rough draft of report.

7. Final report, presentation posters, etc. This should be a few days before the presentation is made in whatever form is selected.

Page 46 is a flowchart summarizing the progress of a project.

Obviously, some of these items will have to be adapted to the grade and development level of the students. Third or fourth graders might be expected to list no more than two or three things they have read, while the high school student might need at least 12-15 references to adequately cover background material. Similarly, the interim progress reports and final report will need to be adapted to the student level. *It is usually inappropriate to set a minimum length for the reports.* If you do, expect that to be all you get. This could be a subject for discussion on an individual basis.

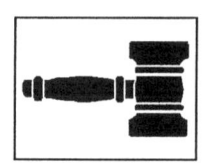

Important note to teachers: be sure to provide parents and students with a copy of the schedule with all due dates.

Expect problems. Make sure that students are aware before they start that most scientific research does not progress without at least one hitch. The problem may or may not be of the student's making. If they know to expect it beforehand, they will be less upset and more able to cope with it when it does occur. If by some miracle it does not occur, the student will be overjoyed!

This is a good place to make younger students aware of Murphy's Law. Older students are well acquainted with it, especially if they have done science projects before.

# Let's Get Started

Despite all efforts of both parents and teachers, most students will not complete their reports and presentations before the last possible minute. This means a hassle to get supplies, usually after most stores have closed for the day. Be prepared ahead of time both at school and at home with the following if they are applicable to the project.

Stockpiling supplies is particularly important for science fair projects which may involve extensive posters and other display materials.

- Glues: white, rubber cement, fabric cement, and epoxy
- Small traveling sewing kit with needle, thread, and an assortment of buttons
- Screw drivers: several sizes of both flathead and Phillips
- Assorted colored felt tip markers
- Portable typewriter
- Dictionary
- Major reference books related to projects which might require quick information or substantiation
- Assorted colored construction paper
- Posterboards
- Batteries

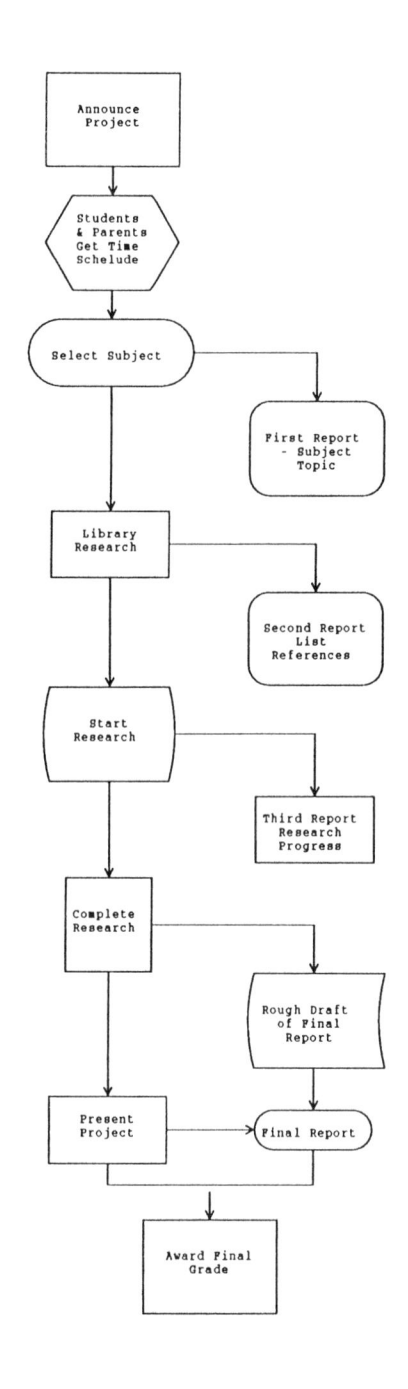

```
        ┌──────────────┐
        │   Announce   │
        │   Project    │
        └──────┬───────┘
               │
        ╱──────┴───────╲
       ╱   Students     ╲
      │   & Parents      │
       ╲  Get Time      ╱
        ╲  Schelude    ╱
         ╲────┬───────╱
              │
        ╭─────┴────────╮
        │ Select Subject├──────────┐
        ╰─────┬────────╯           │
              │                ╭───┴────╮
              │                │ First Report
              │                │ - Subject
              │                │   Topic │
              │                ╰────────╯
        ┌─────┴────────┐
        │   Library    ├──────────┐
        │   Research   │          │
        └─────┬────────┘      ╭───┴────╮
              │               │ Second Report
              │               │    List │
              │               │ References
              │               ╰────────╯
        ┌─────┴────────┐
        │    Start     ├──────────┐
        │   Research   │          │
        └─────┬────────┘      ╭───┴────╮
              │               │ Third Report
              │               │ Research │
              │               │ Progress │
              │               ╰────────╯
        ┌─────┴────────┐
        │  Complete    ├──────────┐
        │  Research    │          │
        └─────┬────────┘      ╭───┴────╮
              │               │ Rough Draft
              │               │ of Final │
              │               │  Report  │
              │               ╰────┬────╯
        ┌─────┴────────┐           │
        │   Present    │    ╭──────┴────╮
        │   Project    ├───▶│Final Report│
        └─────┬────────┘    ╰──────┬────╯
              │                    │
              └────────┬───────────┘
                       │
                ┌──────┴───────┐
                │ Award Final  │
                │    Grade     │
                └──────────────┘
```

# Chapter 10

## Time & Money

The idea of a free education is a myth. Education costs dearly in both time and money for both teachers and parents. Even when classroom supplies are bought with a school purchase order, the cost ultimately falls back upon the taxpayer. The individual special items are an additional expense for parents and often for teachers, too.

During my teaching career, I occasionally let students look at school and laboratory supply catalogs. They were amazed at the costs--not that the costs were unreasonable, but they had simply never thought about it. A little arithmetic quickly gave a sometimes staggering view of how individual item costs become multiplied for a teacher's student load of 100 - 150 students. This turned out to be an impressive item in making students aware of the value of the education they were receiving.

Many parents would be equally amazed if they kept a careful record and added up the total amount spent during a year for various school needs. Many teachers, especially at the elementary level, likewise spend considerable sums of their own money on supplemental materials. Often this totals well over $1,000 during a school year--and few teachers are paid that well. If parents and teachers are not careful, science projects can add substantially to the total outlay.

The question of who picks up the tab is often a difficult

one. The basis for decision is one which must consider the grade level and the type of project required. One approach that I often used is this: if the student is given complete freedom of choice in selecting the project, and if the project produces something which the student wishes to keep, then the costs should be the responsibility of the student. If specific projects are assigned, particularly if they are a curriculum requirement, the costs properly belong in the category of school supplies.

## *Keep Up With Costs*

The additive effect of a dollar or two here and there over a year can become a substantial sum of money. Students as well as teachers and parents need to be aware of costs involved.

A good addition to a science project at any level is a cost analysis. It has a fringe benefit of providing the student an opportunity to do some constructive arithmetic. At the end of this chapter are three convenient charts to use in tracking costs. The first could be provided by either the parents or teacher to help the student in organization of records. Completion of it could actually be a factor in grading the project.

The second chart is a means for parents to keep track of their educational expenditures over a period of time. The third chart provides a convenient means for teachers to be amazed at how much they really spend out of their own pockets during a school year.

# Time & Money

In considering any of these costs, don't overlook the intangibles. Charge the costs of driving your car to obtain supplies at a minimum of twenty cents per mile-- and that cost is very conservative. Time is money--don't forget to add in your time. What do you charge for it? Try even minimum wage--the results of that will look bad enough!

## *Store-bought or home-made?*

When teachers of younger students are preparing for project time, they are often faced with the dilemma of whether to require each student to provide his or her individual supplies, for the teacher to purchase them locally as individual items, or to purchase a kit of supplies from a science education materials source. There are factors in addition to simple dollar amounts that should enter into such a decision. Let's consider an example.

A popular kit used to teach elementary students about gravitropic responses in plants is the "Acrobatic Seedling" by SYNTHEPHYTES™ available from major school supply companies. At the time of publication, this kit had a list price of $22.50. It contains enough supplies for 25 students to work individually. The teacher's manual provides background information and other helps. A copy master for a standard laboratory report is included. Suggestions are made for variations and extensions of the experiment. These could easily be adapted for individualization of projects. Here is a complete list of the kit contents:

Student manuals - 13
Teacher Manual - 1
Lab Report Reproduction Master - 1
Jiffy-7 Instruction Sheet - 1
Plastic Cups - 25
Absorbent Pads - 25
Corn Seeds - 75
Bean Seeds - 75
Jiffy-7 Peat Pellets - 25

Let's see if we can save a few dollars by purchasing the materials individually on a local basis. Your initial problem will be to find a clear container for the experiment. Relatively few suitable ones are available. One is the Solo™ cup which has a price of around $1.09 for a package of 18. You could reduce this cost by using old glass jars or less expensive styrene cups which can shatter easily and cause injury if broken. The few pennies saved would not be the beginning of costs at the courthouse. What about paper towels to take the place of the absorbent pad? Add the cost of paper towels. Two will be required for each student at about 1.5 cents each. Be sure to use white paper towels of good grade. Ascertain that they do not contain a mold inhibitor left over from the manufacturing process because that might interfere with seed germination. Be sure that your seeds are fresh. There is nothing worse than an experiment that does not work and this one will not work if the seeds used do not have a high germination rate. Figure a cost of about eight cents per student for high quality seeds. Don't forget to add in time and car expense spent chasing around trying to find appropriate materials. When you have finished you will probably be able to calculate having saved a few cents per student.

# Time & Money

Is it worth it in this case? What are the chances of the student being disappointed because the paper towels fail to stand up in the cup or would not hold enough moisture to provide for the seeds over the weekend? How much classroom time is going to be wasted distributing and preparing the individual materials? How much additional time will be needed to make out a report form? How much more classroom time will be expended instructing the students what to do? Compare that with the almost none required when you use the kit with its student manuals clearly explaining each step.

The kit has the added bonus of a Jiffy Pellet™ for each student to plant a seed to take home.

Consider one very practical final economic thought. If you buy the kit, the chances are it will be on a school purchase order. If you buy the items locally, chances are, despite your intentions, you will probably never get reimbursed for the costs.

Requiring the students to obtain their own supplies is not a very sensible alternative either. Each will have to purchase the full amount contained in a standard package size, so in effect the cost calculated above would be multiplied by 25 with much of the surplus material going to waste. That is not good education.

The above is not meant to imply by any means that professionally prepared and packaged supplies are always a desirable answer. In many cases, greater educational objectives are accomplished in project work if the students' own ingenuity must provide materials from common or even scrap sources. If the projects are to be exhibited in some type of competition such as a science fair, the judges will often put more weight on the use of homemade materials than they will on purchased kit materials.

In summary, yes, project work is going to cost money. But so does just about everything else.

# Project Cost Data

Project Title _____    Student _____

| Date | Material |  | Transportation |  |  |  |  |
|------|----------|------|----------------|---|---|---|---|
|      | Name | Cost | Miles X $.20 = Cost |  |  |  |  |
|      |      |      |                |   |   |   |   |
|      |      |      |                |   |   |   |   |
|      |      |      |                |   |   |   |   |
|      |      |      |                |   |   |   |   |
|      | Total |     |                |   |   |   | Total |
|      |      |      |                |   |   |   |   |

Total Cost (Materials + Transportation) $ _____

## Parents School Cost Summary

| Material |        |      | Transportation         |            |
|----------|--------|------|------------------------|------------|
| Date     | Name   | Cost | Miles X $.20 = Cost    | Time Spent |
|          |        |      |                        |            |
|          |        |      |                        |            |
|          |        |      |                        |            |
|          |        |      |                        |            |
|          |        |      |                        |            |
|          |        |      |                        |            |
| Total    |        |      | Total                  |            |

## Teacher Expenses

| Date | Class | Material |  | Transportation |  |
|------|-------|----------|--|----------------|--|
|      |       | Name | Cost | Miles X $.20 = Cost | Time Spent |
|      |       |      |      |      |      |
|      |       |      |      |      |      |
|      |       |      |      |      |      |
|      |       |      |      |      |      |
|      |       |      |      |      |      |
|      |       |      |      |      |      |
|      |       | Total |  | Total |  |

# Chapter 11

## Let's Do A Project

At this point you may be thinking something like "All this sounds real great, but does it work in real life? How do you do really do it?"

All right, let's try it. Here is an example of how you can take an ordinary subject which really is not scientific and turn it into a scientific investigation encompassing all of the attributes of good research. This should be fun. It will also produce something edible--at least part of it should be! The project is based on making bread. The title will be "Aunt Julia's Bread" because the recipe is a real one from the author's great aunt, Julia Oswald (1888-1989). We understand that she obtained the recipe from her mother so it goes back at least to the mid-1800's. As you will see, this is a basic recipe which could be varied to produce different types of bread products. Somewhere along the line some of the original "add so and so until it looks right" got converted into some actual measured quantities. The recipe is given on the next page.

# Let's Do A Project

## Aunt Julia's Bread (Small Batch)

Measure into sauce pan:
- 1 1/2 cup milk
- 1 tablespoon (large, rounded serving spoon) shortening
- 1/4 cup sugar (1/2 cup for cinnamon rolls)
- 1 tablespoon (measuring spoon) salt

Heat together and let cool until lukewarm.

In large glass or measuring cup:
- 1 yeast cake, broken up
- 1/2 cup lukewarm water
- 1 teaspoon sugar

Let set until it comes to top of cup. Stir until dissolved. Pour into sauce pan.

Add:

- 7 cups flour, or more
- 1 egg

Mix all together. Kneed on floured board. Put in greased bowl and let rise until double in bulk. Punch down. Kneed again. Shape into loaves, rolls, or roll out for cinnamon rolls. (For cinnamon rolls--roll out, dot with butter, sprinkle with sugar and cinnamon. Roll up and slice.) Let rise and bake in moderately hot oven (375°) until brown and bread has a hollow sound when thumped.

Now, let's see how this simple kitchen recipe could be turned into scientific research. Several questions about it come to mind. These could be worked out together to produce a large extensive project, or each question in itself might serve as the basis for a simpler project.

1. **What is the effect of varying the quantity of the flour used?** The quantity of flour used in the original recipe produces a fairly heavy, dense bread. By using less flour Aunt Julia obtained rolls, cinnamon rolls, and other goodies. Get the "less flour" to measurable quantities. Use the same basic recipe but divide it into several portions. Start off with a portion that has just enough flour to make the dough barely solid. Use different increments of increase to obtain the original quantity. Figure 11.1 shows the dough with minimal flour. Figure 11.2 shows it as prepared according to the recipe. Cook all the different flour concentrations at the same time in the same type and size of pan. When completed, weigh each for comparison. Another comparison relating to texture would be to measure and calculate an average size for the holes produced in the bread. (These probably have some formal name that would be given them by the great chefs of the world, but I frankly haven't the faintest idea.) Anyway, you get the idea. The ultimate test would be an organoleptic test. In biology, that means any test based on a subjective analysis by use of the senses. In other words, now is the time to eat some!

# Let's Do A Project

**Figure 11.1**

**Figure 11.2**

2. **What is the effect of the sugar concentration?** For this one, keep the amount of flour and everything constant in the various portions, but vary the amount of sugar from none through several increments to the amount specified in the recipe. What effect does this have on the bread rising, and ultimately on its texture? This would be a good place to send them to the library to learn something about fermentation processes. Be prepared for questions about what happens to the alcohol produced. Yes, the bread fermentation does yield some alcohol which is volatile and boils off during cooking. Maybe that's why some people like raw bread dough.

3. **What effect does temperature have on the time required for the bread to rise?** Make up a batch of dough and let it go through the fermentation (rising) process at different temperatures. Comparc time to rise to the cooking point with the temperature of incubation. Another question that might be asked in conjunction with this is what effects, if any, the different temperatures have on the final texture?

4. Another project could be based on the number of times the dough is kneaded and allowed to rise. **Would variation here produce a different quality final product?**

5. **What might be the effects of different cooking temperatures?** This recipe probably originated a century before kitchen ranges became equipped

# Let's Do A Project

with thermostats. How did they judge it?

6. Another investigation could be based on cooking the bread in an open top pan as compared with cooking it in one covered with aluminum foil. **What differences might result?**

7. **What effect does the size of the cooking pan have?** In other words, would you obtain the same texture if the bread were cooked in a large as compared with a small pan?

8. **Is there a difference in time required or other factors related to the type of yeast used?** Compare a yeast cake with the freeze-dried powdered form of yeast.

9. **What kind of flour should be used, or would it be desirable to use a different kind for different types of bread products?** A trip to the grocery store quickly reveals an intimidating number of different types of flour, to say nothing of brands. Which should be chosen? You have recently seen one promoted by its producer as being "Best for homemade bread." What is the difference between this flour and other varieties produced by the same maker? Why would the difference be important?

10. ... and on and on and on... By this time you have the idea. At this point the first batch should be ready for the initial organoleptic test.

## Protocols

Every scientific experiment should have a written plan for the research. Such a plan is called a **protocol**. There is no set absolute form for one, but it should cover all the major steps. The next page gives one form of protocol for bread experiment 1. The same procedure is shown in a flow chart set-up.

### Let's Do A Project

## *Protocol For Bread Experiment #1*

Note: All portions should be baked in identical pans.

1. Mix together all ingredients except for flour as directed in recipe.

2. Add two cups of flour to mixture and stir thoroughly.

3. Remove one-fourth of the dough and place in a pan or bowl.

4. Add 1 cup of flour to remaining mixtures and stir in.

5. Remove one-third of the dough and place in a pan or bowl.

6. Add 1 cup of flour to remaining mixture and stir in.

7. Remove one-half of the dough and place in a pan or bowl.

8. Add 1 cup of flour to remaining mixture and stir in.

9. Place in pan.

10. Follow directions for rising and kneading.

11. Bake according to directions.

12. Measure height of finished loaves.

13. Compare weight of finished loaves.

14. Cut each loaf in half and measure 10 air spaces in each to obtain average size.

15. Apply organoleptic test to a small portion of each loaf.

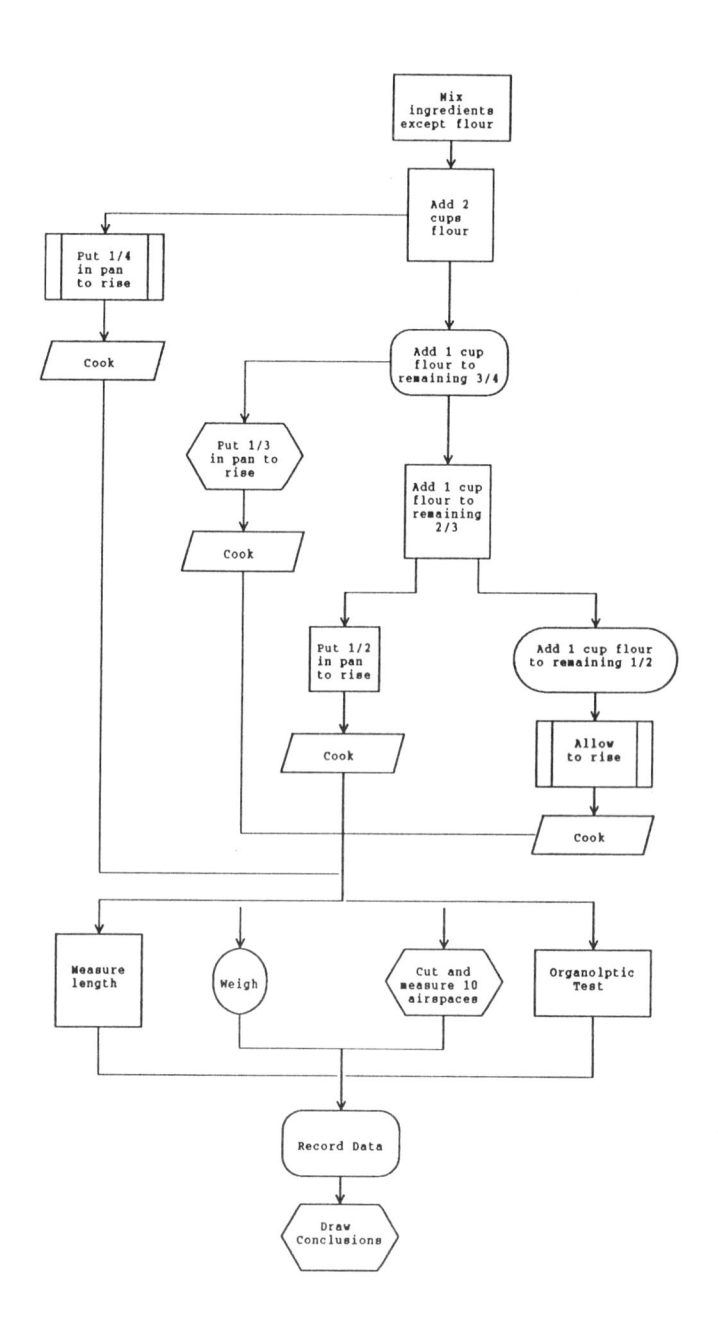

```
                    ┌──────────────┐
                    │     Mix      │
                    │  ingredients │
                    │ except flour │
                    └──────┬───────┘
                           ▼
                    ┌──────────────┐
         ┌──────────│    Add 2     │
         │          │    cups      │
         │          │    flour     │
         │          └──────┬───────┘
         ▼                 │
  ┌────────────┐           │
  │  Put 1/4   │           │
  │  in pan    │           │
  │  to rise   │           │
  └─────┬──────┘           │
        ▼                  ▼
    ╱Cook ╲         ┌──────────────┐
                    │  Add 1 cup   │
                    │  flour to    │
          ┌─────────│ remaining 3/4│
          │         └──────┬───────┘
          ▼                │
      ⬡Put 1/3⬡            │
       in pan to           │
         rise              ▼
          │         ┌──────────────┐
          ▼         │  Add 1 cup   │
        ╱Cook╲      │  flour to    │
                    │  remaining   │
                    │     2/3      │
                    └──────┬───────┘
              ┌────────────┴──────────┐
              ▼                       ▼
      ┌────────────┐          ⬡Add 1 cup flour⬡
      │  Put 1/2   │          ⬡to remaining 1/2⬡
      │  in pan    │                  │
      │  to rise   │                  ▼
      └─────┬──────┘             ┌────────┐
            ▼                    │ Allow  │
          ╱Cook╲                 │to rise │
                                 └───┬────┘
                                     ▼
                                   ╱Cook╲
```

```
  ┌──────────┐    ○        ⬡Cut and⬡    ┌──────────┐
  │ Measure  │  Weigh      ⬡measure 10⬡ │Organolptic│
  │  length  │             ⬡airspaces⬡  │   Test   │
  └─────┬────┘    │            │        └─────┬────┘
        │         │            │              │
        └─────────┴──────┬─────┴──────────────┘
                         ▼
                    ╱ Record ╲
                    ╲  Data  ╱
                         │
                         ▼
                    ⬡  Draw  ⬡
                    ⬡Conclusions⬡
```

# Let's Do A Project

## Related Topics

Now it's time to tie the science project into other educational disciplines. The most obvious would be mathematics. Get involved with measurement. For students who have enough math background, calculate averages and other means of comparisons.

Obviously, English comes into play in preparing the report on the project. Insist that everything be grammatically correct and that all words are properly spelled.

This one could even be tied into history. Students might be required to include something about the history of bread making as background information in their project report.

Art could also be involved. What would happen if you did a little sculpture work with the bread dough when it is placed in the pan prior to cooking!

One of the greatest things about a good project idea is that it leads to others. Let your imagination go. Think of related cooking or kitchen topics which could be developed as we did this one. With just a little bit of brain-storming the possibilities become almost endless! Who said project ideas are hard to find? Some times we don't see the forest..., or is it we don't see the trees?

## Data And Reports

Remember that a vital part of any science project is keeping adequate records. You will find it easy to make a variation of one of the data forms used in many lab manuals for a specific project. Or better still, discuss what is needed with the students and have them devise their own data recording chart. Complexity, of course, would depend upon the student's grade level. You can see how easy it would be to change the headings in the basic forms setup to provide for any of the experiments suggested in this book.

## *In Conclusion*

That wasn't so bad was it? The important thing here is that successful science projects do not have to be boring and difficult. They can be fun for all concerned. They do not have to utilize a lot of sophisticated equipment and expensive materials. They do require some preparation and thinking.

Obviously, the type of experiment described above is not one which is suitable for an honors high school senior who wants to use the project for science fair and progress all the way to the International Fair final judging. Those students do not need the kind of help we're talking about here, although hopefully someone years earlier did provide it and got them started off on the career track they are now following. What a satisfaction that is for the earlier teachers!

# Appendix A

## Project Books

Bonnet, Robert L. & G. Daniel Keen. 1991 Botany-49 Science Fair Projects. Tab Books. Blue Ridge Summit, PA.

----- -----. 1991 Botany-49 More Science Fair Projects. Tab Books. Blue Ridge Summit, PA. 1991.

Gardner, Robert. 1986 Ideas for Science Fair Projects-Experimental Science Series Book. Franklin Watts, N.Y.

Iritz, Maxine Haren. 1991 Science Fair-Developing a Successful and Fun Project. Tab Books. Blue Ridge Summit, PA.

Loiry, Williams S. 1983 Winning With Science. Loiry Publishing House.

Tant, Carl, 1991. Seeds, etc... Biotech Publishing, Angleton, TX.

Tant, Carl, 1992. Science Fair Spelled W-I-N. Biotech Publishing, Angleton, TX.

Tocci, Salvatore. 1986 How to do a Science Fair Project. Franklin Watts, N.Y.

# Appendix B

## Science Supply Sources

Benz Scientific
P.O. Box 7022
Ann Arbor 48107
(313) 994-3880

Connecticut Valley Biological
Supply Company
P.O. Box 326
Southampton, Mass. 01073
(413) 527-4030

Edmund Scientific Co.
101 East Gloucester Pike
Barrington, NJ 08007-1380
(609) 573-6250

Flinn Scientific
P.O. Box 219
Batavia, IL 60510-9906
(708) 879-6900

Nasco
901 Janesville Avenue
Ft. Atkinson, WI 53538
1-800-558-9595

Nebraska Scientific
3823 Leavenworth Street
Omaha, Nebraska 68105
1-800-228-7117

Sargent-Welch Scientific
7400 North Linder
Skokie, IL 60076
1-800-727-4368

Sargent-Welch Scientific, Canada
77 Enterprise Drive North
London, Ontario N6N 1A5
1-800-265-3496

Science Kit
& Boreal Laboratories
777 East Park Drive
Tonawanda, NY 14150-6784
1-800-828-7777

Starr Education Products
P.O. Box 3077
Athabasca, Alberta
Canada TOG OBO
(403) 675-5510

Synthephytes
P.O. Box 1032
Angleton, TX 77516-1032
(713) 369-2044

# Index

assignments . . . . . . . . . . . . . . . . . . . . . . . . . . . . . 11
classroom . . . . . . . . . . . . . . . . . . 29, 33-35, 37, 47, 51
costs . . . . . . . . . . . . . . . . . . . . . . . . . . . . . . 29, 47-51
data . . . . . . . . . . . . . . . . . . . 6, 10, 14, 28, 41, 65, 66
experiments . . . . . . . . . . . . . . . . . 24-26, 29, 35, 36, 66
grading . . . . . . . . . . . . . . . . . . . . . . . . . . . . 20, 34, 48
help . . . . . . . . . . . 2-4, 9-12, 15, 19, 22, 37, 38, 48, 66
judges . . . . . . . . . . . . . . . . . . . . . . . . . . . 37-39, 41, 52
kits . . . . . . . . . . . . . . . . . . . . . . . . . . . . . . . . . . . . 29
parents . 2, 4, 9-11, 18-20, 22, 32-35, 37, 38, 42, 44, 45
records . . . . . . . . . . . . . . . . . . . . . . . 20, 23, 41, 48, 65
schedules . . . . . . . . . . . . . . . . . . . . . . . . . . . . . . . . 33
science fair . 4, 24, 28, 31, 35, 37, 38, 40, 45, 52, 66, 67
stress . . . . . . . . . . . . . . . . . . . . . . . . . . . . . 3, 4, 18, 43
supplies . . . . . . . . . . . . . . . . . . . . . . 29, 45, 47-49, 51
teacher . . . . . . 4, 8-10, 19-21, 29, 30, 33, 39, 42, 47-50
time . . 1, 2, 12, 19, 20, 22, 25, 29-34, 40, 42, 45, 47-51

## Colophon

Text entry and layout format were done with **WordPerfect**™ **5.1** by Amy Harris and Renee' Walker.

**Synthephytes** is a trademark of Plant Something Different, Inc.

**WordPerfect** is a trademark of WordPerfect Corp.

**Solo**™ **Cups** is a trademark of Solo Cup Company.

Printed in the USA by
BookMasters, Inc.
Ashland, Ohio